Water Water Everywhere: But Not a Drop to Drink

William (Bill) C. McElroy
Copyright Jan 2021
Updated April 2023

ISBN: 9798598968406

Preface:

Water Water Everywhere, but not a drop to drink refers to water that has been so polluted that it is no longer potable or even good for other uses. The author has watched our environment undergo major changes during his lifetime, and access to clean pollution free water has and is continuing to be a challenge for many of our planet's citizens.

This short manuscript contains dozens of Internet link addresses as references that you can use to further your curiosity about our water and its future. This manual is small in length, but if you access all the Internet pages referenced, you will find hundreds of pages of valuable information on our plight, that of obtaining fresh potable water for all our world's citizens.

The links provided were tested and are good as of March 2023. This links were selected for minimal commercial content and maximum valuable information. The author recommends reading each of the web pages, but unfortunately has no control over the links or content, or if the owners of the website pages will continue to post each in the future.

Please note that the title is actually a play on the phrase from the following:

(1798) A poem by Samuel Taylor **Coleridge** about an old sailor who is compelled to tell strangers about the supernatural adventures that befell him at sea after he killed an albatross, a friendly sea bird. A famous line is "**Water, water, everywhere, / Nor any drop to drink**." The speaker, a sailor on a becalmed ship, is surrounded by salt water that he cannot drink.

Table of Contents:

Water Water Everywhere: But Not a Drop to Drink 1
Preface: 1
Table of Contents: 3
Chapter #01 – The Farm: 5
    Section # 01 – The Hand-Dug Well 5
    Section # 02 – From Where did the Water Come: 6
    Section # 03 – The Drilled Wells: 6
Chapter #02 – Some History of Water: 8
    Section # 01 – Farmers of the Old West: 8
    Section # 02 – River and Lake Use 9
    Section # 03 – First Water Purification Plants: 9
    Section # 04 – First Long-Distance Pipelines and Aqueducts: 10
    Section # 05 – History of Water and Codes: 11
Chapter # 03 – Chemical Pollution: 11
    Section # 01 – Natural Earth Chemicals 12
    Section # 02 – Industrial Chemicals: 12
    Section # 03 – Pharmaceutical Chemicals: 13
    Section # 04 – Airborne Chemicals: 14
    Section # 05 – Oil and Gas Drilling: 15
Chapter # 04 – Waterborne Diseases: 17
    Section # 01 – Pandemics: 17
    Section # 02 – History of Waterborne Diseases: 17
    Section # 03 – The Diseases that Sicken or Kill: 17
Chapter # 05 – Dams and Electric Power: 19
    Section # 01 – Hoover Dam, and Others: 19
    Section # 02 – Damage to the Environment: 19
    Section # 03 – Damage to the Wildlife: 20
    Section # 04 – Silt Buildup: 21
    Section # 05 – Dangerous Dams: 21
Chapter # 06 – Bottled Water and Its Negative Effects: 22
    Section # 01 – Why We need Bottled Water: 22
    Section # 02 – Cost of Bottled Water: 22
    Section # 03 – City and Townships: 23
Chapter #07 – From Where do We Obtain Our Water?: 24

Section # 01 – Surface and Groundwater: 24
Section # 02 – Overuse of Water: 25
Chapter # 08 – The Future of Water: 27
Section # 01 – Filth: 27
Section # 02 – Lack of Water: 27
Section # 03 – Cost of Water: 28
Chapter # 09 – Solutions to Water's Demise: 29
Section # 01 – Pipelines: 29
Section # 02 – Recycled Water: 29
Section # 03 – Hydrogen Generation 31
Section # 04 – Reverse Osmosis: 32
Section # 05 – Seawater Desalting: 32
Section # 06 – Cloud Seeding: 33
Section # 07 – Conservation: 33
Chapter # 10 – Possibilities: 37
Section # 01 – Mountain Valley Towns: 37
Section # 02 – Turbine Powered Pipelines: 37
Section # 03 – Drilling & Pump Limits: 38
Section # 04 – Mandatory Reclaim Systems: 38
Section # 05 – Water Rationing: 38
Section # 06 – Better Food Production: 38
Section # 07 – Hot Water Recirculation Systems: 39
Section # 08 – Catch Basins: 39
Section # 09 – Storage Reservoirs: 39
Section # 10 – Limit Use of Fossil Fuels: 40
Section # 11 – Atmospheric Water Generation: 40
Chapter # 11 - Abbreviations & Laws: 42
Index: 43
Author: 48
Cover Picture: 48

Chapter #01 – The Farm:

The author's parents moved from Brooklyn, New York to a farm in the Catskills in 1944. The farm had close to 100 acres of very poor land, and there was no electricity, no roads, no natural gas, no inside toilets, and no septic system. It did have a hand-dug well, a three-hole outhouse, a small swamp, one or two fields for growing hay, a very large two-story garage, and a large barn. The building was built in the early to mid 1800's and was put together with hand-hewn beams held together with wooden pegs. The floors were rough-sawed logs over a foot wide and with bark intact. Heat was from a Pot-Belly stove, and cooking was on a two-burner wood stove. We had 32 milk cows, 36 cats, 1 dog, several chickens, four to six pigs, and lots of snakes, woodchucks, squirrels, deer, skunk, mice, and other assorted products of nature. Dad was in the US Army and I was two-years old at the time, mother and I ran the farm most of the time. I was hunting squirrel with a Model 06 Winchester, driving the tractor, chopping wood, and doing other chores at the age of five. It was a wonderful life, clean air, lots of wild fruit and vegetables, lots of fish and game, and miles of land to explore.

Section # 01 – The Hand-Dug Well

It was on the farm that I first became familiar with water and its value to life. We had an eighteen-foot deep hand-dug well for our primary water supply. There was a creek that ran wet during winter, spring, and part of summer, and it was used for washing clothes, and bodies. Cold as ice, but it kept us clean.

The well contained about two and a half foot of water at its bottom, and we dipped a bucket on a hoist into it, and pulled the precious liquid to the surface. We did eventually add a hand pump. It was used to water the livestock, make our meals, take baths when the creek was dry, and water the vegetable garden. Surprisingly it never got higher than two and a half foot high, and very seldom got lower. Our neighbors that lived on the hill above us would come down during the heat of the summer and get water for their survival; the well provided for all.

The water in the well was some of the best tasting water I have every remembered drinking. And I have drunk water from nations around the world and just about every state in the U.S. It was pure, sweet, cool, and delicious.

Section # 02 – From Where did the Water Come:

The farm was at the foot of the Shawangunk Mountains and we were directly downhill from Mud Pond one of several lakes that dot the Shawangunk Ridge. The lakes were fed by rain and there were few if any entry or exit points; water left by filtering down through the bottom until it reached a layer of solid rock. As it turns out, the rock layer acted as a conduit for the water to flow underground for several miles and some of that filtered water found its way to our Hand-dug Well.

Section # 03 – The Drilled Wells:

Over the decades from the 1940s to the 2020s the area has built up with homes and small businesses, and the demand on water grew along with this growth. The Hand-dug well era had ended, and the drilling machines took over. We had sold the farm years prior to the 1970s, but kept one lot for ourselves, and in 1972 built on this lot. The well drilled had to go down about 200 feet until it hit water. The water was still from the mountain lakes, and still tasted pretty good, but the water in the nearby stream became unsafe to use do to the many homes with septic tanks, leach fields, and livestock that came to being in the interim.

Naturally, human nature took its turn with a hint of greed. Others purchased two of the adjacent wooded, and swampy lots, and they immediately sought water. The one owner drilled some 200 or so feet and hit the same water supply as we had, and thus no real problem. The second owner drilled 200 feet, decided it would be better to drill several more hundred feet and hit a layer of Sulfur contaminated water, which naturally rose to contaminate all the wells in the area, including ours. This therefore required a new storage system be installed that was designed to dispel the Sulfur and its rotten egg smell.

Others up the mountain from us also drilled deep wells that hit minerals and other contamination, and therefore the once clean, clear, delightful water of the 1940s was forever a thing of the past.

Chapter #02 – Some History of Water:

I am sure that if we could talk to some cavemen and women that they would tell us about their search for clean potable water. Our Native Americans in the Southwest not only depended on water, but also managed to find ways of storing it. Some tribes walked for days to get water and bring it back to the village; and unfortunately in some sections of our planet we still have this situation, i.e. parts of Africa and many of our southwestern Native American areas.

If you go back to the days of the various empires you will find that the citizens developed ingenious methods of water storage, water transport, water filtering, and water diversions. Stone aqueducts that channeled life-giving water from mountain lakes to far-away towns and cities were built all over Europe, China, and the Middle East. Some of these water channels can still be seen today, and many are now tourist attractions.

A good website for reading about the history of water (2021) is https://www.lenntech.com/history-water-treatment.htm

This site takes you back to 2000 BC and describes some of the techniques that were used over the centuries for filtering out contaminants.

Section # 01 – Farmers of the Old West:

When our western territories were occupied there were many disputes over the ownership of water from rivers, lakes, ponds, creeks, and springs. If you owned the property on which the water existed, you therefore claimed to own the water, and the right to keep or sell it as you pleased. This did not go well with settlers and especially some farmers; and some of the disputes not only ended up in lawsuits, but also in the death of one or more of the people involved.

Section # 02 – River and Lake Use

Water has a tendency to flow downhill to eventually settle in our oceans. Without some sort of natural or man-made dam, there would be no lakes. Lakes tend to allow pollution to settle to the bottom and thus the water just under the surface may be reasonably potable. For centuries people used lake water for their potable water without realizing that animals, fish, and humans excrete feces and urine into the water. Additionally, water that is not flowing tends to harbor bacteria and other pathogens that are dangerous to human life.
REF: https://www.epa.gov/environmental-topics/water-topics

As civilization advances it uses pesticides and fertilizers for growing edible crops for animal and human consumption. We, the citizens of our world, also tend to discard all sorts of garbage on our streets and elsewhere. We also use powerboats for transportation, sport, and fun on our lakes and the engines used may add oil and other petroleum byproducts to the water.
REF: https://www.treehugger.com/lake-pollution-types-sources-and-solutions-1204112

Section # 03 – First Water Purification Plants:

Purifying water to make it palatable for drinking or bathing has been the goal of humans for thousands of years. All sorts of filtering, boiling, distilling, and chemical means have been tried. A civil engineer named Robert Thom established the first municipal water purifying plant in 1804 in Scotland. It took 93 years before the Maidstone, England city managed to chlorinate their entire water supply. In 1827 the first sand filter was created and used by an Englishman named James Simpson. Jersey City, New Jersey was the first U.S. city to use disinfectants in their community water; the year was 1908.
REF: https://www.cdc.gov/healthywater/drinking/history.html

Section # 04 – First Long-Distance Pipelines and Aqueducts:

Around 2000 BC the Chinese used reeds for the transport of water from streams, lakes, and rivers to their populated areas. It was in the 1800s that modern steel pipes came about for carrying water. In the U.S. cities like New York City in 1842 used aqueducts and pipes to bring water to the population. Other cities like Philadelphia and Seattle used hollowed out logs for water distribution. It was about 1850 when laws were passed that required filtering systems for the transported water; mostly due to bacteria contamination that made people very sick or worse, dead.
REF: https://en.wikipedia.org/wiki/New_York_City_water_supply_system

A good history of Aqueducts in early Europe, i.e. by the Romans in 312 BC can be read at the following:
REF: https://www.britannica.com/technology/aqueduct-engineering
REF: https://www.thoughtco.com/aqueducts-water-supply-sewers-ancient-rome-117076

The future may bring us long-distance water pipelines, but do not hold your breath as the current Chinese Pipeline is in trouble, both dollar wise and time wise. There are also many questions that have to be answered by those financing and building a long-distance water pipeline. But, if water becomes too scarce, we may have to bite the bullet and go for it.
REF: https://www.lizsilverman.com/2015/03/long-distance-water-pipelines-possibility/

Section # 05 – History of Water and Codes:

The history of water use and its attempts to collect and regulate goes back for many centuries; this link provides a condensed look at that history.
REF: https://www.iveyengineering.com/historical-events-plumbing-systems/

Chapter # 03 – Chemical Pollution:

Water is a caustic that will eat into most natural and man-made substances. It is great for cleaning, for cooling, for heating, for bathing, and for drinking. Since it can dissolve many substances the chemicals that make up those substances can easily contaminate it. As water passes through the earth flowing toward the oceans, it comes in contact with dirt, rock, and all the minerals within each.

Many animals die in the wild and their remains become a health hazard to humans and other animals. Rain and the discharge of biological materials onto the earth or into the streams, lakes, etc., can pollute the water that eventually is part of our potable water systems.

Dead plants and garbage from human activity can and does frequently contaminate our water supplies. Storms that create fast moving water flows can stir up the bottom silt which may contain fecal matter and all sorts of decay.
REF: https://www.nrdc.org/about

Minerals that may be harmful to humans can be dissolved in our lakes, rivers, streams, and aquifers, thus contaminating our water.

In some coastal areas like southern Florida the groundwater is being replaced by salty seawater as the oceans rise due to Global Warming, and also from too much removal of fresh water from the subsurface aquifers.

REF: https://www.usgs.gov/mission-areas/water-resources/science/saltwater-intrusion?qt-science_center_objects=0#qt-science_center_objects

Section # 01 – Natural Earth Chemicals

The soil around us contains all sorts of minerals, bacteria, and chemicals that can be both beneficial and harmful to human life. We have our oceans and our landmasses that collect all sorts of man-made and natural components. Many of these are created from chemical reactions caused by decomposition or by intense heat of wildfires or volcanoes. Many of these are soluble in water, and thus are part of our water supply and therefore, may be toxic to life.

Section # 02 – Industrial Chemicals:

Over the decades since around 1900 we have been in the Industrial Age where manufacturing boomed and jobs abounded. What was not known, or maybe it was, is that many of the industrial processes produced chemicals that were poisonous to life on our planet.

If a manufacturing or chemical company uses water for its processes, the water may become contaminated with the chemicals being used. Many of these chemicals are toxic and hazardous to one's health.

Ford Motor Company in Mahwah, New Jersey for years was dumping toxins in the Ramapo Mountains (Ringwood) and it was not until the Mahwah plant closed that it was fully made public. The toxic materials were contaminating the Ramapo River and much of the local water supply. Ford ended up paying millions to clean up the mess that they had made.
REF: https://www.epa.gov/enforcement/settlement-addresses-groundwater-cleanup-ringwood-mineslandfill-superfund-site-new

The Passaic River in New Jersey was one of the worst examples of industrial chemical pollution in that there were several

companies dumping Arsenic, Lead Batteries, Agent Orange, Dioxins, Pesticides, and other chemicals in the river. This happened over two centuries before the EPA and other environmental groups put an end to the practice.
REF: https://response.restoration.noaa.gov/about/media/how-do-you-begin-clean-century-pollution-new-jerseys-passaic-river.html

The Animus River pollution was a man-made disaster caused by the EPA in its effort to clean up another man-made pollution disaster. The EPA workers were attempting to drain water from an abandoned Gold Mine when the soil holding back the water separated and flooded millions of gallons of pollution into the Animus. The pollution was a mix of Lead, Arsenic, and other toxins, and eventually made its way into the Colorado River a major source of potable water for millions of residents in several states.
REF: https://www.epa.gov/goldkingmine

Section # 03 – Pharmaceutical Chemicals:

Do you pee? I trust that your answer will be 'Yes'. The human body during its lifetime on this planet may and usually does get sick at one time or another. Our medical and pharmaceutical industry has learned to cash in on the fear of being sick or possibly dying and therefore has developed not hundreds, but thousands of chemicals that may be beneficial to stopping pain and sickness.

As we take these chemicals into our bodies, we use the properties, and then we tend to excrete each via feces and urine. This urine is then flushed into our sewer or septic systems and eventually ends up in our future water supply. People also stop using their medications and many are disposed of by flushing down the toilet or sink drains, a situation that should be avoided.

Municipal water companies do their best to eliminate these chemicals by treating or filtering the water, but it is nearly impossible to eliminate 100%.

SEE: *Performance Data for the Clearly Filtered Inline Fridge Filter* – It is scary to think that all 250+ contaminations could be in our drinking water.

Section # 04 – Airborne Chemicals:

We for decades spread all sorts of chemicals and fertilizers on our lawns, farmlands, and lakes for the purpose of increasing crop yield and reducing the growth of parasites. Some of these chemicals were eventually eliminated due to water pollution from rain runoff. The problem is that as soil dries it may become loose and thus, be blown by the wind for dozens to hundreds of miles. This airborne chemically polluted dust then settles down on our land, vehicle, and buildings, only to once again be washed by rain into our potable water supplies.
REF: https://www.indianaenvironmentalreporter.org/posts/indiana-university-researchers-monitor-airborne-pollution-in-great-lakes

Have you ever visited Mono Lake in California? I ask because several years ago the lake that contains all sorts of harmful chemicals was partly drained, and the dried up lake bottom soil was picked up by the wind with much of it being deposited over southern California. Since that lesson, the lake has been kept reasonably full.
REF: https://deeply.thenewhumanitarian.org/water/articles/2016/04/27/mono-lake-facing-another-crisis

AQI

AQI stands for the Air Quality Index, a measure of the cleanliness of the air we breathe. It can be calculated using the following formula:

"The Air Quality Index is based on measurement of particulate matter (PM2.5 and PM10), Ozone (O3), Nitrogen Dioxide (NO2), Sulfur Dioxide (SO2) and Carbon Monoxide (CO) emissions." "All measurements are based on hourly readings: For instance, an AQI reported at 8AM means that the measurement was done from 7AM to 8AM."
REF: https://waqi.info/

Section # 05 – Oil and Gas Drilling:

The oil and natural gas industries are important to the current survival of the U.S., and much of the world. The majority of our buildings are heated, the majority of our transportation is moving, and the majority of our manufacturing depends on the low cost of crude oil and natural gas products.

The problem is that the industry for over a hundred years did not conduct itself in a manner that environmentalists consider necessary to protect our earth, air, and water. Oil spills, the use of water for Fracking, and the hydrocarbon pollution from burning fossil fuels has a proven record of destroying life-giving forest, water supplies, and human health.
REF: http://blogs.edf.org/energyexchange/2017/12/29/six-ways-oil-and-gas-development-can-contaminate-land-and-water-and-what-to-do-about-it/

To get oil and gas out of the ground is not an easy process, there are not large swimming pools of crude oil under the ocean or land, but a mix of porous rock that contains millions of droplets or molecules of oil or gas. To get these to the surface the companies have to pump water into the rock and thus, allow the water to displace the oil and gas, which is then captured via pipes and brought to the surface for processing.
REF: https://www.filtsep.com/water-and-wastewater/features/oil-and-gas-water-treatment-in-oil-and-gas/

Much of the Gulf of Mexico is considered a 'Dead Zone' (Hypoxic Zone) from pesticide, fertilizer, and leaking oil well

pollution. There is next to no life in a dead zone, no plants, no fish, and little to no bacteria. The water is not fit for human, plant, fish, or animal life.
REF: https://oceantoday.noaa.gov/happenowdeadzone/

Chapter # 04 – Waterborne Diseases:

Waterborne diseases account for nearly 80% of the diseases that sicken or kill people around the world.

Section # 01 – Pandemics:

Water from rivers, lakes, streams, reservoirs, swimming pools, distribution systems, and in-building piping has over the decades caused pandemics or near pandemics due to viruses and bacterium polluting the source or the storage. The following link describes the many pollutions and the consequences of each. REF: https://www.doh.wa.gov/Portals/1/Documents/5100/420-044-Guideline-WaterOutbreak.pdf

Section # 02 – History of Waterborne Diseases:

One of the first examples of known waterborne disease was in 1854 in London, England where sewage polluted water caused a Cholera outbreak that sickened or killed thousands. New York City in the 1800s had several outbreaks of sickness and due to this decided to create a citywide water distribution system. REF: https://www.nap.edu/read/11010/chapter/3#18

Section # 03 – The Diseases that Sicken or Kill:

Cholera, Dysentery, Escherichia Coli, Giardia, Hepatitis A, Typhoid, and Salmonella are the most common diseases that can be transmitted by water that has been polluted by items like animal or human feces. There are many more diseases and causes that are waterborne and International Health Organizations are continuously working to eliminate each.

REF: https://lifewater.org/blog/7-most-common-waterborne-diseases-and-how-to-prevent-them/

Chapter # 05 – Dams and Electric Power:

The idea behind a dam is that it stores water for future use, it can provide energy for generating electricity, it provides a recreational paradise, and it helps prevent catastrophic floods. The building of a dam provides work for dozens to hundreds of citizens, and many dams double as roadways across otherwise impossible to cross rivers. Dams come in all sizes, shapes, from a simple beaver dam on a small creek to a mile wide dam across a major river. Many dams are fitted with flood control spillways, electric turbine generators, fish ladders, canal gates, and a visitor's center for tourists.
REF: https://www.ussdams.org/dam-levee-education/overview/benefits-of-dams-levees/

Section # 01 – Hoover Dam, and Others:

FDR, Franklin Delano Roosevelt needed a way to end the Great Depression of the 1930s and one of his means was to hire out of work citizens to build a great dam across the Colorado River. The project was one of the largest implemented and when completed became a power and water source for not just Las Vegas, but for several southwestern states, including much of southern California.
REF: https://rarehistoricalphotos.com/building-hoover-dam/

Section # 02 – Damage to the Environment:

Many dams were built on rivers that provided protection to the environment in the event of heavy rains. The flowing river water moved fast enough to keep sensitive wooded, grasslands, farmlands, etc., from being flooded. The flowing water also carried junk, garbage, fallen trees and branches, dead animals, etc. downstream to the areas more adapt to handling each. A dam slows the water flow, causes water back up, and causes trash and other items to collect, thus further blocking the flow.

The fallen matter can settle to the bottom of the reservoir and then rot in the stilled waters; this rotting can cause methane and other greenhouse gasses to develop and percolate into the atmosphere, thus adding to our Global Warming problems.

Recently it has been found that dams are causing some earthquakes. The weight of the water in the reservoir slowly compresses the earth below it, and this causes movement on the side areas of the dam that eventually fail, thus an earthquake.
REF: https://greentumble.com/how-dams-affect-the-environment/

Section # 03 – Damage to the Wildlife:

The open flow of water from its source to the oceans is beneficial to many breeds of fish, especially the Salmon that need to spawn in the place where they were born. Salmon in the rivers are food for not just humans, but also for bears, birds, and many other animals, and therefore if you block the river with a dam, the food supply may become limited or non-existent.

Stagnant water, like that found at the bottom of a deep reservoir can cause plant life, and sunken garbage to rot and this can cause Mercury to change state into Methylmercury, which is a poison to fish and humans. This poison can contaminate the meat of the fish that are eaten by the local animals, and therefore poison them.

The silt buildup can also cover rocks that fish use for their hiding place or homes, and it can choke out the needed oxygen that many fish and plants need.
REF: https://www.americanrivers.org/threats-solutions/restoring-damaged-rivers/how-dams-damage-rivers/

Section # 04 – Silt Buildup:

One thing that most people do not know is that dams block silt, the fine dirt that is pulled downstream by the river's current. Once the silt hits the calm water near the dam it settles to the bottom of the reservoir and slowly over the years builds up and up. This buildup will eventually need to be removed or the dam will become useless. The silt can collect to the level of the dam's outlets, and this now can cause all sorts of floating junk and trees to jam up the outlet. The silt buildup also depletes the amount of water that is available from the reservoir.

Removing the silt can be a major problem since millions of cubic feet of silt would have to be sucked out, dredged, or otherwise flushed. Flushing every ten or more years means draining most of the reservoir in hours, thus releasing its waters into the river valleys below the dam. Doing this can ruin the environment and also put people living downstream in danger. To pump out or dredge the silt, means hauling it away and finding a place to deposit it; not the easiest feat.
REF: https://www.hcn.org/articles/water-as-sediment-builds-a-colorado-dam-faces-its-comeuppance-paonia-reservoir

Section # 05 – Dangerous Dams:

Since the building of the Hoover Dam the world has built hundreds to thousands of wood, dirt, rock, tailings, and concrete dams on just about every river in the world. The estimate in 2021 is that there are some 800,000 known dams. The problem is that many are well over their original design life age and now have become a danger to people and wildlife that are living downstream from each.
REF: https://www.motherjones.com/environment/2019/07/the-dam-truth-the-91000-dams-in-the-us-earned-a-d-for-safety/

Chapter # 06 – Bottled Water and Its Negative Effects:

In the last few decades there has been a move away from using well and municipal water for drinking and cooking. The 'sales pitch' is that bottled water is cleaner, purer, and more healthy than the water supplied by personal wells or municipal water supplies.

Section # 01 – Why We need Bottled Water:

It is true that some well water and some municipal water does contain chemicals, minerals, and other pollutants. Most homeowners do not have their well water tested on a regular basis and do so only when the water turns brown, or smells, or is making household members sickly. Municipal water in small towns may not be treated for all the possible contaminants, and municipal water in large cities may be treated but still has to travel miles through age-old pipelines that may have leaks, dead water spots (at the end of some streets), etc.

Due to the fear of contaminated water we tend to trust and purchase bottled water, especially that labeled as 'Pure Spring Water'. Well, spring water sounds great, except that most of the springs in today's world are contaminated with all sorts of nasty stuff, due to being surrounded by human and animal development.

Section # 02 – Cost of Bottled Water:

In 2019 the citizenship of the United States spent an estimated $18.5 Billion dollars on bottled water, and produced millions of tons of plastic waste from the discarded bottles. It also cost for removal of the bottles and the cleanup of the bottles from our roads, streets, lakes, rivers, streams, and oceans.

The cost of obtaining the water, bottling the water, marketing the water, selling the water, and the cost of the oil used for the process is quite high, and if you really check, the water is no safer to drink than most tap water. In fact, the bottles used for the water can leach harmful chemicals into the so-called pure water.
REF https://www.jerseyislandholidays.com/plastic-bottle-pollution-statistics/

Section # 03 – City and Townships:

Much of the bottled water that is sold is actually tap water from municipal supplies, and the companies that are producing this 'Pure' bottled water are in fact conning the public. Much of the bottled water tested by authorities contains the same amounts of contamination that is found in municipal water. Additionally, many towns have had there wells run dry due to over-pumping by the bottle water companies, that drain the local aquifers and then move to another town, draining theirs.
REF:
https://www.theguardian.com/environment/2019/oct/29/the-fight-over-water-how-nestle-dries-up-us-creeks-to-sell-water-in-plastic-bottles

Chapter #07 – From Where do We Obtain Our Water?:

Simple question, complex answer. Water is a combination of two elements; that being Hydrogen and Oxygen, the formula is H2O, which translates to two parts Hydrogen and one part Oxygen. Hydrogen and Oxygen in nature are gaseous substances and when combined with intense heat produce water.
REF: https://sciencenotes.org/making-water-from-hydrogen-and-oxygen/

Section # 01 – Surface and Groundwater:

Our oceans contain most of the earth's water supply, but due to runoff water from our land masses each contains all sorts of minerals, including salt. Salt water is great for cooking and preserving, but is not so great for drinking and therefore, ocean water is not a good source of potable water in its current state.
REF: https://www.dumblittleman.com/drinking-salt-water/

Fortunately, the sun is continuously sending us rays of light and heat that changes ocean water into saturated cloud water, and our high and cooler mountains change this cloud water into rainwater. The rainwater then becomes surface water as it collects in streams, puddles, rivers, lakes, etc. Water has a unique feature in that it tries to seek the lowest levels, which on earth are our oceans. Thus, water runs downhill to the lowest level that can contain it, i.e., an underground aquifer or the ocean. As the water seeks its lowest point the sand and other materials in the soil filter it. This filtering action produces somewhat pure water that can be pumped to the surface and used by homeowners and municipalities for their water supply.
REF: https://www.usgs.gov/special-topic/water-science-school/science/freshwater-withdrawals-united-states?qt-science_center_objects=0#qt-science_center_objects

Throughout the world, even in what seems to be very dry desert areas, there are aquifers that contain millions of gallons of water. We in 2021 have satellites that can look beneigh our topsoil and see the water and thus provide an estimate of the amount that is still available. Unfortunately, some of the studies show alarming overuse of the subsurface water that may take decades to replenish.
REF: http://www.geologypage.com/2018/07/scientists-use-satellites-to-measure-vital-underground-water-resources.html

Section # 02 – Overuse of Water:

To some in areas of the world where there seems to be an abundance of water, it is impossible to use too much of it. Unfortunately, we have to eat, and to eat we need to grow things; things like animals and plants. In the U.S. we have an abundance of food available due to our farmers and those farmers from other nations. Farming is a Federally Subsidized activity and thus, many farms grow things just to get the money, and then they bury or destroy the crops, or byproducts such as milk. Farming uses up to 70% of all the available water and for the most part much of the water is wasted due to evaporation, over watering, and system leakages.
REF: https://www.wired.com/2006/03/farms-waste-much-of-worlds-water/

Farming animals produces a need for clean water for cleaning the hides, the meat, and the machinery used for dissecting the animals in preparation for market. This too uses much of the available water that may be very scarce. In addition to this, if raising animals in barn then there is the cleanup of animal feces and urine, and this can use up to 150 gallons per animal per day.
REF: https://www.onegreenplanet.org/environment/shocking-facts-on-how-factory-farms-cause-water-pollution/

We also have the farms that are located in areas of the nation where each should not be. For example, Beef and dairy, pecan, rice, grape, and cotton farming, in states like Arizona where water is very scarce and must be channeled into the state from

hundreds of miles distant, i.e. the Colorado River. There are aquifers under some of the deserts, but well drilling by agricultural corporations is fast lowering the water to dangerous levels.
REF: https://www.azcentral.com/in-depth/news/local/arizona-environment/2019/12/05/biggest-water-users-arizona-farms-keep-drilling-deeper/3937582002/

In the Central Valley of California we have cotton and raisin grape productions that are consuming millions of gallons of northern California water. In the populated areas of the state there is an overuse of landscaping water, and it too must eventually be scaled back by better conservation and different landscaping practices.
REF: https://www.ppic.org/publication/water-use-in-california/

Chapter # 08 – The Future of Water:

As the earth grows older the numbers of people living on it grows and grows. At present, 2021, we have about 6.3 billion people and many times that in animals, and each requires a daily dose of clean, pollution free, water.

Section # 01 – Filth:

It is unfortunate that as we have more people, we have more pollution, and that pollution, be it in the ground or air somehow enters our potable water supplies. We generate millions of tons of trash each year, and this trash is dumped into our oceans or into landfills. When it rains, the trash is oxidized or otherwise broken down and forms rust, dirt, and other materials that contain toxins, many of which are harmful to the environment and to life forms. We have tried to eliminate many of these by burning our trash, or mixing it with other chemicals, or filtering, or letting it settle out, but none of these techniques fully eliminate the filth that pollutes.

Section # 02 – Lack of Water:

We have oceans and we have glaciers and ice fields, and each of these produce water that can be used for sustaining life and our industrial base. As the climate warms up the glaciers and ice fields are melting and will eventually be lost as a source of water. Many streams, rivers, and lakes will therefore dry up as the source disappears. This in turn will dry up the aquifers that provide water to many towns and cities.

Currently, the change in climate, 2020, has created massive drought conditions in the west and American Southwest, and this lack of water is causing problems for farmers and for urban citizens. Water rationing is on the way, if not already being done in some townships.

Section # 03 – Cost of Water:

The cost of water may be one of the lowest cost utilities or owner costs for many companies, renters, and homeowners. There was a time when to help manufacturing, the water companies reduced the cost per unit (ccf) (748 Gallons) of water if you used more and more of it. This has changed, and today if you use over set limits the cost to you may go up and up.

We should be using all sorts of water saving techniques in our homes and businesses, and in all new construction. Yes, it may cost an extra $1,000 to the cost of building a home, but over a 30-year mortgage at say 5%, the $ 1,000 is only $5.37 per month. The monthly savings in the cost of the water can easily meet or exceed that amount.

The true cost of water can vary from area to areas, from usage to usage, and from the amount of money needed to purify it. Water cost can range from pennies per gallon to dollars per gallon, and unfortunately some citizens use precious water to manufacture products or grow things in areas where water is scarce and should not be used.
REF: https://www.greenbiz.com/article/true-cost-water

Water is a commodity just like gas, oil, gold, food, etc., and like it or not, it can be sold for a profit or traded on the stock market. This means that although we may have plentiful water today, we may not tomorrow, and the investors can and probably will jack up prices and place many citizens out of the range of buying water.
REF: https://www.bloomberg.com/news/articles/2020-12-06/water-futures-to-start-trading-amid-growing-fears-of-scarcity

Chapter # 09 – Solutions to Water's Demise:

Gold is expensive, about $1,800 or so per ounce. Gasoline is still reasonable at under $4.50 per gallon. Water though in the future will be more expensive than both if we do not change are wasteful ways, conserve it, and stop polluting it.

Section # 01 – Pipelines:

It amazes me that we can and do build tens of thousands of miles of crude oil, natural gas, and other petroleum pipelines that crisscross the nation over rivers, mountains, deserts, grasslands, swamps, etc. Yet, we do not seem to be able to build pipelines from natural water sources like the Great Lakes or the thousands of northern lakes to the desert southwest or the Native American Reservations.
REF: https://www.bridgemi.com/michigan-environment-watch/how-long-can-great-lakes-fend-thirsty-world-water-diversions

It boils down to money and greed in that our priorities are for fuel for our vehicles, home and office heating, and manufacturing factories, rather than water for the survival of humans that are water deprived.

Yes, we do have some pipelines like those that bring water over the mountains into Southern California and those that bring water into New York City, but these are relatively short in length, and mostly paid for with taxpayer dollars via delivery fees.
REF: https://www.nrdc.org/sites/default/files/Water-Pipelines-report.pdf

Section # 02 – Recycled Water:

One way to conserve water is to not use it, but that is not possible in today's world, so we need other means, as such recycling.

Grey Water:

There is a current push by building code authorities to add water reclaim to new home construction. This 'grey' water reclaim system is a means of saving the water that has soap in it from washing dishes, clothing, and one's body. The idea is to have a separate drain system that can detect the amount of pollution in the wastewater and direct it to the on-site storage tank, or if too polluted to the city or townships waste water, i.e., sewer system.

The homeowner for watering plants, cleaning sidewalks and driveways, etc, then uses the gray water.
REF: https://elemental.green/complete-beginner-guide-to-greywater-systems/

Urine Reclaimed Water:

Some will tell you that it is ok to drink one's urine in the event of severe thirst and no other water being available. Do not do it. NASA (National Aeronautics and Space Agency) has worked for decades to come up with a safe way to create drinking water out of urine when humans are aboard their space vehicles. The cost of the existing machine is $250 million and it has not yet proved to be reliable. Fortunately, or unfortunately since this was not developed in the U.S., some foreign companies are working on how to filter urine for use in third world countries where potable water is scarce. It has also been found that urine can be used to generate electricity that can be used in homes. NASA is considering the system for space travel as well.
REF: https://www.globalcitizen.org/en/content/water-filter-astronaut-urine-global-water-crisis/

Rainwater Capture:

Cities like Tucson, Arizona are offering dollar rewards for installing rainwater collection systems. Homeowner's must follow certain rules for the collection of the rainwater and have a means of using it for irrigation, cleaning, and other non-potable uses.

REF: https://www.cdc.gov/healthywater/drinking/private/rainwater-collection.html

The systems collect rainwater from your roof and direct it to a storage container for future use, or to areas of lawn and garden that need the extra moisture. Filters and a pump system may be needed if the storage tank is not elevated enough for gravity flow of the water when needed.
REF: https://www.instructables.com/How-to-build-a-rain-water-collector/

Section # 03 – Hydrogen Generation

There are several means of generating hydrogen, all of which have a fairly high cost. Hydrogen when burned with Oxygen produces one by-product and that is water or H2O. Hydrogen therefore is an excellent substitute for gasoline, fuel oil, natural gas, coal, wood, and other combustibles currently being used for generating heat or power.
https://www.pv-magazine.com/2020/06/22/low-cost-direct-solar-to-hydrogen-ambitions-see-the-light/

The problems with hydrogen, other than the cost to produce, are the safety of storage and the cost of the machines that can be used to burn it. Hydrogen is very active and explosive when mixed with Oxygen and therefore has to be tempered with other items. It can and should be metered so as to not produce explosions that will destroy the engines that are designed to use it.

Hydrogen Powered Vehicles can rival gasoline powered vehicles in power, distance, and negative pollution. Several companies are experimenting with hydrogen-powered vehicles and some with water to hydrogen powered vehicles.
REF: https://www.popularmechanics.com/cars/hybrid-electric/a22688627/hydrogen-fuel-cell-cars/

31

Section # 04 – Reverse Osmosis:

If you do an Internet search for Reverse Osmosis Filters you will find dozens for sale at hardware and lumber yards. The systems are for home use and contain several types of filters, and a means for storing the cleaned water.

Reverse Osmosis filtering removes most minerals, dirt, and other contamination, but does not remove bacteria, parasites.
REF: https://watertechadvice.com/how-reverse-osmosis-works/

To remove bacteria and parasites a UV light is used. UV tends to kill bacteria and some claim the COVID-19 virus.
REF: https://www.nationalacademies.org/based-on-science/covid-19-does-ultraviolet-light-kill-the-coronavirus

Section # 05 – Seawater Desalting:

Taking seawater and using it to flood small ponds that use the sun to evaporate it is a technique for mining sea salt. Several companies do this in places like the Bay Area of California. There is nothing to stop us from pumping seawater to desert areas of the nation and allowing the super hot sun to evaporate it, thus leaving behind sea salt and several other types of minerals.

To use seawater as potable water suitable for drinking, cooking, and bathing, we have to use different techniques. One such technique is distilling, which is similar to allowing the sun evaporate the water, but this time we use heat to evaporate the water and then we capture the condensate. The salt and minerals are left behind, and the condensate when cooled condenses into reasonably pure water. This process uses a lot of costly fuel, but can be accomplished with low cost Green Energy.

The second process is to force pump the seawater through special filters that remove most of the salt and minerals, then distill the near pure water.
REF: https://www.aqueum.com/water-treatment/unit-processes/desalination/

The major objection to using seawater is that the pumping process can also pump sea creatures, small fish, and plankton into the system, thus possibly changing the environment of the area from which the water was obtained.
REF: https://www.scientificamerican.com/article/why-dont-we-get-our-drinking-water-from-the-ocean/

Section # 06 – Cloud Seeding:

One possible means of generating more water is to seed the clouds so that each produces rain. Cloud Seeding is expensive, as it requires an aircraft to fly close to or above the clouds as it dumps silver iodide, dry ice, potassium chloride, or table salt into the clouds. This helps to cool the clouds to minus 40 F or more and these therefore caused the clouds to release water in the form of rain, ice, or snow.

The process is being used in very dry desert areas and elsewhere. The process works best in areas that actually have sufficient rain or snow, as it adds about 10% to the normal down flow of water. This allows for filling lakes, rivers, and reservoirs much faster and with fuller.
REF: https://www.treehugger.com/what-is-cloud-seeding-4863907

Section # 07 – Conservation:

One of the best means of saving water is to just not use it. By cutting down on the use of water we take some of the load off of our filtration plants, pipelines, aqueducts, aquifers, and reservoirs. The average household of four people uses about 400 gallons of water per day. There are about 120,000,000 households in the U.S., and therefore we use close to 48,000,000,000 gallons of water per day. If we can cut this by just 10%, we can save 4,800,000,000 gallons of water each day. That fills a lot of swimming pools.

Residential Water Conservation:

The use of low flow toilets and showers; water efficient dishwashers and clothes washers, and the prevention of leaking pipes and valves can go a long way in obtaining that 10% savings. Some cities like Tucson, Arizona have even offered free low flow toilets to low-income residents, and tax breaks to others if they agree to replace their old toilets, shower heads, and water wasting appliances.
REF: https://www.tucsonaz.gov/water/low-income-high-efficiency-toilet-rebate

Cities like Los Angeles have over the decades grown in population, but have due to water conservation and other practices actually cut water use.
REF: http://projects.scpr.org/applications/monthly-water-use/los-angeles-department-of-water-and-power/

Cities like Tucson, Arizona has water conservation codes that residents must adhere to: See the listing and links to each code at:
REF: https://www.tucsonaz.gov/water/ordinances#:~:text=For%20more%20than%2025%20years%2C%20Tucson%20Water%20has,active%20and%20enforceable%20for%20many%20years%20to%20come.

Other ways to conserve water are to flush less. The saying is "If it's Yellow, let it mellow; if it's Brown, flush it down". You should also only do full loads of dishes or laundry, and you should sweep, rather than hose, sidewalks and driveways when possible. You can also save water by selecting plants and grasses that are native to the area in which you live.

Commercial Water Conservation:

Here is the thing though; most of our available water actually is used on commercial farms to grow our food. Agricultural water use has been increasing to an alarming rate, especially in drought

prone areas of our nation. Some parts of Arizona and the Central Valley of California are slowly tapping out their water supplies due to poor use of existing water, poor selection of crops being grown, and poor means of reclaiming.
REF: https://www.cdc.gov/healthywater/other/agricultural/index.html

There are currently active programs to cut farm use of water by 2% or more. These include high-tech water of crops, selecting crops that use less water, using drip irrigation, using U.S. government satellite data for analyzing crop fields and water usage, and sharing information between each farm.

REF: https://www.unilever.com/sustainable-living/reducing-environmental-impact/water-use/working-with-our-suppliers-and-farmers-to-manage-water-use/

Commercial and Medical Building Water Conservation:

Like a home, commercial buildings, including hospitals and medical buildings, use water for urinals, toilets, showers, cleaning, laundry, drinking, and other uses. Therefore, each should comply with applicable building codes, the U.S. Green Building Council, and with the National Energy Policy Act. Low flow toilets, pressure regulators, insulated hot water pipes, grey water reclaim, rainwater harvesting, and proper inspection and maintenance of all cooling towers, piping, and valves can save millions of gallons of water.
REF: https://www.csemag.com/articles/10-ways-to-save-water-in-commercial-buildings/

Commercial Manufacturing Water Conservation:

Manufacturing plants should use much of the same water conservation as residential and commercial buildings. In addition to these conservation measures, the plants should also look into means of recycling all the water used in the actual manufacturing

processes, if allowed to by applicable laws and sanitary conditions.
REF:
https://www.worldbank.org/content/dam/Worldbank/Feature%20Story/SDN/Water/events/IWREC2014-Session2A-Water-Use-Manufacturing-RandyBecker-Sept8.pdf

Chapter # 10 – Possibilities:

In the future we will need to use all of the conservation and sanitation techniques referenced in this manual as well as new techniques that have yet to be attempted. Some of these are harsh and there will be much objection by those that feel our governments are too powerful. The fact is that we are heading for a major catastrophe in many areas of the world due to the lack of pure and safe drinking, cooking, and bathing water. Here are some ideas to consider:

Section # 01 – Mountain Valley Towns:

Water in the sun will evaporate and most reservoirs are fully open to the sun. In the west we have many mountainous areas that contain deep valleys. We should consider building our towns and cities on platforms over these valleys and using the bottom of the valleys for clean water storage, i.e. a shielded reservoir. The reservoirs can and will collect rainwater, and can also be used for pumped in water or using existing stream or river water.

Section # 02 – Turbine Powered Pipelines:

The pipelines that feed water to New York City and to Los Angeles have steep drops that provide lots of pressure to the water as it travels to its destination. Unfortunately, in the Los Angeles case, and in some other city and town pipelines there has to be strong electric pumps to push the water up hills and over flatlands.

Think about this, we can install turbines in the water lines and use the force of the water flow to generate electricity; basically free electricity that can be used to not only power the pumps, but also be sold over the power networks.

Section # 03 – Drilling & Pump Limits:

This is harsh, but it may be needed to save many farms, towns, and other municipal water supplies. Limit the depth of the water drilling and pump to 25% of the known aquifer water depth. So if the top of the water in an aquifer is at 500 feet and the bottom of the water table is at 1,000 feet, then the maximum drill depth would be 750 feet. This would give access to the water between 500 and 750 feet and no more. This does two things, it provides for future water and it helps to prevent sinkholes and earthquakes caused by removal of the water that is supporting the ground above it.

Section # 04 – Mandatory Reclaim Systems:

New housing and commercial buildings may in the future be mandated to include water recirculation, recycled, grey water separation, water harvesting, and insulated water systems that use most of the existing and future water conservation technology. This will increase the building cost and selling or renting prices, but will cut the cost of water use over the years, thus a possible net zero in total cost.

Section # 05 – Water Rationing:

This is a situation we hope to avoid, but if we continue to have droughts, excessive water usage, or excessive depletion of our aquifers and reservoirs, then water rationing may be mandatory and controlled by water meters that automatically reduce water flow when excessive amounts of water is used.

Section # 06 – Better Food Production:

Much of our water is used for food production and a considerable amount of our produced food goes to waste due to over farming, poor transportation, spoilage, and a host of other situations. We may need to eliminate or reduce the availability of many crops and animals, and we may have to find better means of farming that limit the use of the existing water. This is already on the way to being practiced, but needs to be massively expanded around the world as our population keeps increasing.

Section # 07 – Hot Water Recirculation Systems:

Most households waste hundreds to thousands of gallons of water each year due to poorly designed water heating and distribution within the buildings. The owners run the water until all the cold water is out of the pipes, and then they collect the hot water for drinking, cooking, bathing, shaving, washing dishes or clothes, etc. Mandatory insulated recirculation systems will add to the cost of building, but will save water and its associated cost. We need this change included in our building codes, and building permits not only for residential buildings, but also for all new buildings.

REF: https://www.epa.gov/sites/production/files/2017-01/documents/ws-homes-hot-water-distribution-guide.pdf

Section # 08 – Catch Basins:

Rain and snow that falls on buildings, streets, sidewalks, lawns, fields, etc., mostly goes to waste as it flows toward our oceans. Some cities and towns are now creating catch basins that can collect this water for distribution use. These catch basins can be separate from the general population, but better yet available as ponds in parks that are stocked with fish and can be used for non-motorized boating or swimming.

Section # 09 – Storage Reservoirs:

If you have been to Tucson, Arizona you will see on the hillside a large windowless building. This is actually an underground reservoir that contains a rubberized bladder that stores the drinking water for thousands. Being covered, it has very little waste due to evaporation, and it is protected from external pollution from dust, plants, birds, and other animals

REF: https://www.waterworld.com/drinking-water/article/16193028/flexibility-key-to-underground-water-reservoir-rehabilitation

Many areas of our world receive rain or snow intermittently, and this intermittent access to water is collected in open ponds, or reservoirs where evaporation takes its toll. Perhaps an investment in more enclosed and insulated storage should be considered.

Section # 10 – Limit Use of Fossil Fuels:

As seen in another area of this manual, fossil fuels like coal, oil, and natural gas are currently necessary, but in the future may not, nor should be. We have many alternative energy sources that can save valuable water from being polluted by crude oil and its by-products. Solar, wind, wave, bio, nuclear, geothermal, hydroelectric, hydrogen, and other sources of power and heat are slowly being accepted by our population as the cost is reduced and the availability is increasing. More homes are being lighted or heated by solar panels; more vehicles are using electric motors; and more public transportation is using hydrogen for fuel.

REF: https://www.scientificamerican.com/article/renewable-energy-saves-water-and-creates-jobs/

Section # 11 – Atmospheric Water Generation:

These devices already exist, and you may even be able to make one at home. The idea is to extract pure drinking water from the air. The way it works is to cool down the moist air to the point that the moisture condenses and collects in a storage container.

We all take baths, wash and shave, flush toilets, wash dishes and clothing, drink water, cook with water, and wash floors with water. This water can partly saturate the air in your home along with the normal moisture you let into the home every time you open a door or window. The generator extracts and captures this water for use at a minimum cost to you.

REF: https://www.epa.gov/sites/production/files/2019-11/documents/awg_technical_brief_final_05nov19.pdf

Chapter # 11 - Abbreviations & Laws:

This is not all-inclusive as there may be thousands of Federal, State, and Local as well as foreign laws that govern how our environment and water supplies are protected from disease and other pollution.

Air Quality Index (AQI)
Ambient Water Quality Standards (AWQSs)
Atmospheric Water Generator (AWG)
Beaches Environmental Assessment and Coastal Health (BEACH) Act of 2000
Central Arizona Project (CAP) canal.
Centum (Hundred) Cubic Feet (CCF)
Clean Water Act of 1977 (CWA)
Enhanced Surface Water Treatment Rule (LT1ESWTR)
Federal Water Pollution Control Act (FWPCA)
Ground Water Rule (GWR)
Interim Enhanced Surface Water Treatment Rule (IESWTR)
National Pollutant Discharge Elimination System (NPDES)
National Research Council (NRC)
Net Present Value (NPV)
Public Water Systems (PWSs)
Safe Drinking Water Act (SDWA)
Surface Water Treatment Rule (SWTR)
Total Coliform Rule (TCR) and the
U.S. Environmental Protection Agency (EPA)
Water-enabled Electricity Generation (WEG) Technologies
Water Quality Act of 1987 (WQA)

Index:

2000 BC 8, 10
312 BC 10
748 Gallons 28
Abbreviations & Laws 42
Agent Orange 13
Agricultural Water 34
Air Quality Index 14
Airborne Chemically
  Polluted Dust 14
Airborne Chemicals 14
Alternative Energy 40
American Southwest 27
Animal Feces 25
Animus River 13
AQI 14
Aqueducts 10
Aquifer 38
Aquifers 27
Arizona 25, 35
Arsenic 13
Atmospheric Water
  Generation 40
Author 48
Bacteria 9, 32
Bacteria Contamination 10
Bacterium Pollution 17
Bay Area 32
Better Food Production 38
Biological Materials 11
Bottled Water 22
Building Codes 35, 39
Building Cost 38
Building Permits 39
California Water 26
Carbon Monoxide 15

Catch Basins 39
Catskills 5
ccf 28
Central Valley 26, 35
Chemical Pollution 11
Chinese 10
Chinese Pipeline 10
Cholera 17
City and Townships 23
Cloud Seeding 33
Colorado River 13, 19, 26
Commercial and Medical
  Building Conservation .. 35
Commercial Manufacturing
  Conservation 35
Commercial Water
  Conservation 34
Conservation 33
Conservation and Sanitation
  Techniques 37
Conservation Codes 34
Cost of Bottled Water: 22
Cost of Obtaining Water ... 23
Cost of Water 28
Cover Picture 48
COVID-19 Virus 32
Creek Water 5
Crop Yield 14
Crude Oil 15
Cut Farm Use of Water 35
Damage to the Environment:
  19
Damage to the Wildlife 20
Dams 19
Dams are Fitted With 19

43

Dams double as Roadways 19
Dams Provide Protection ..19
Dangerous Dams 21
Dead Plants and Garbage..11
Dead Zone 15
Deep Valleys 37
Design Life of a Dam 21
Diseases that Sicken or Kill 17
Disinfectants 9
Distilling 32
Drilled Wells 6
Drilling & Pump Limits....38
Drought Conditions 27
Droughts 38
Dry Desert Areas 25
Dysentery.........................17
Earthquakes 20, 38
Electric Power 19
Electric Pumps.................37
Electric Turbine Generators 19
Eliminating Water Pollution 27
Escherichia Coli...............17
Europe.............................10
Farmers of the Old West: ...8
Farming.....................25, 38
Farming Animals 25
Farms 25
FDR 19
Fear of Contaminated Water 22
Fecal Matter....................11
Feces and Urine 9
Federally Subsidized 25
Fertilizers 9, 14

Filtering Action 24
Filtering Seawater............32
Filtering Systems 10
Filth.................................27
First Sand Filter 9
Fish Ladders 19
Florida's Groundwater......11
Flowing Water 19
Ford Motor Company 12
Foreign Companies...........30
Fracking..........................15
Franklin Delano Roosevelt 19
Free Low Flow Toilets 34
From Where......................6
Future Water....................38
Giardia 17
Glaciers...........................27
Global Warming 11, 20
Gold Mine.......................13
Great Depression 19
Great Lakes.....................29
Green Energy...................32
Greenhouse Gasses...........20
Grey Water 30
Gulf of Mexico 15
H2O 24, 31
Hand-Dug Well 5, 6
Hepatitis A......................17
History of Aqueducts........10
History of Water 8
History of Water and Codes 11
History of Waterborne Diseases 17
Hollowed Out Logs 10
Hoover Dam 19, 21
Hose Use..........................34

44

Hot Water Recirculation
  Systems 39
Household Water Use 33
Hydrocarbon Pollution 15
Hydrogen 31
Hydrogen Generation 31
Hydrogen Powered Vehicles
  31
Hypoxic Zone 15
Ice Fields 27
If it's Yellow 34
Index 43
Industrial Age 12
Industrial Chemicals 12
International Health
  Organizations 17
Internet Link Addresses 1
James Simpson 9
Jersey City 9
Lack of Water 27
Lakes Tend To 9
Landscaping 26
Las Vegas 19
Laundry 34
Leaking Pipes 34
Limit Use of Fossil Fuels . 40
London 17
Long-Distance Pipelines ... 10
Los Angeles 34, 37
Low Flow Toilets 34
Mahwah, New Jersey 12
Maidstone, England 9
Mandatory Reclaim Systems
  38
Maximum Drill Depth 38
Mercury 20
Methylmercury 20
Mining Sea Salt 32

Moist Air 40
Mono Lake 14
Mountain Valley Towns ... 37
Mountainous Areas 37
Mud Pond 6
Municipal Water 22
NASA 30
National Aeronautics and
  Space Agency 30
National Energy Policy Act
  35
Native American
  Reservations 29
Native Americans 8
Native Plants 34
Natural Earth Chemicals ... 12
Negative Effects 22
New Home Construction .. 30
New York City 10, 17, 37
Nitrogen Dioxide 15
Oceans 24
Oil and Gas Drilling 15
Oil Spills 15
On-site Storage Tank 30
Over-Pumping 23
Overuse of Water 25
Oxygen 20, 24, 31
Ozone 15
Pandemics 17
Parasites 14, 32
Passaic River 12
Pathogens 9
Pesticides 9
Petroleum Byproducts 9
Pharmaceutical Chemicals 13
Pharmaceutical Industry ... 13
Philadelphia 10
Pipelines 29

45

Plastic Waste 22	Separate Drain System 30
Possibilities 37	Septic Tanks 6
Potable Water Supplies 27	Sharing Information 35
Pot-Belly Stove 5	Shawangunk Mountains 6
Powerboats 9	Shawangunk Ridge 6
Problems With Hydrogen . 31	Silt Buildup 20, 21
Pure Spring Water 22	Sinkholes 38
Rainwater 24, 37	Solutions to Water's Demise
Rainwater Capture 30	 29
Rainwater Collection Systems 30	Stagnant Water 20
	Stone Aqueducts 8
Ramapo Mountains 12	Storage Reservoirs 39
Recycled Water 29	Storms 11
Recycling 35	Sulfur Contaminated Water 7
Removing Silt 21	Sulfur Dioxide 15
Reservoirs 37	Sun's Rays 24
Residential Water Conservation 34	Surface and Groundwater . 24
	Table of Contents 3
Reverse Osmosis 32	The Future of Water 27
Ringwood 12	Trash 27
River and Lake Use 9	Tucson 30, 34, 39
Robert Thom 9	Turbine Powered Pipelines
Romans 10	 37
Rotten Egg Smell 7	Turbines 37
Runoff 24	Typhoid 17
Sales Pitch' 22	U.S. Green Building Council
Salmon 20	 35
Salmonella 17	Underground Aquifer 24
Salt water 24	Underground Reservoir 39
Samuel Taylor Coleridge .. 2	Underground Water 6
Satellite Data 35	Unsafe Water 6
Satellites 25	Urine 30
Save Millions of Gallons .. 35	Urine Reclaimed Water: ... 30
Sea Water Desalting: 32	UV Llight 32
Seattle 10	Water Efficient Dishwashers
Seawater Pumping Process 33	 34
	Water Heating and Distribution 39
Selling or Renting Prices .. 38	

Water in the Future 29
Water is a Caustic 11
Water is Great for 11
Water Meters 38
Water Purification Plants 9
Water Rationing 27, 38
Water Saving Techniques . 28

Water to Hydrogen Vehicles
.... 31
Waterborne Diseases 17
Well Water Testing 22
Why we need Bottled Water
.... 22

Author:

The author has been involved in the environment for many years of his life. He spent a year as an Oceanographer, most of a year at Mercury Nevada studying atomic radiation, and many years traveling the country viewing the mistakes and successes that have affected our environment. His other easy to read books on the environment, travel, history, and self-improvement are on Amazon under the name William (Bill) C. McElroy.

Cover Picture:

Lake Mohave, California.
Lake-Mohave_2-copy.JPG

www.ingramcontent.com/pod-product-compliance
Lightning Source LLC
Chambersburg PA
CBHW050315220526
45465CB00005B/2008